28 Kindergarten Math Puzzles

Practice various Kindergarten math skills by putting the puzzles together.

Shapes, number recognition, counting, addition, subtraction, & more are included!

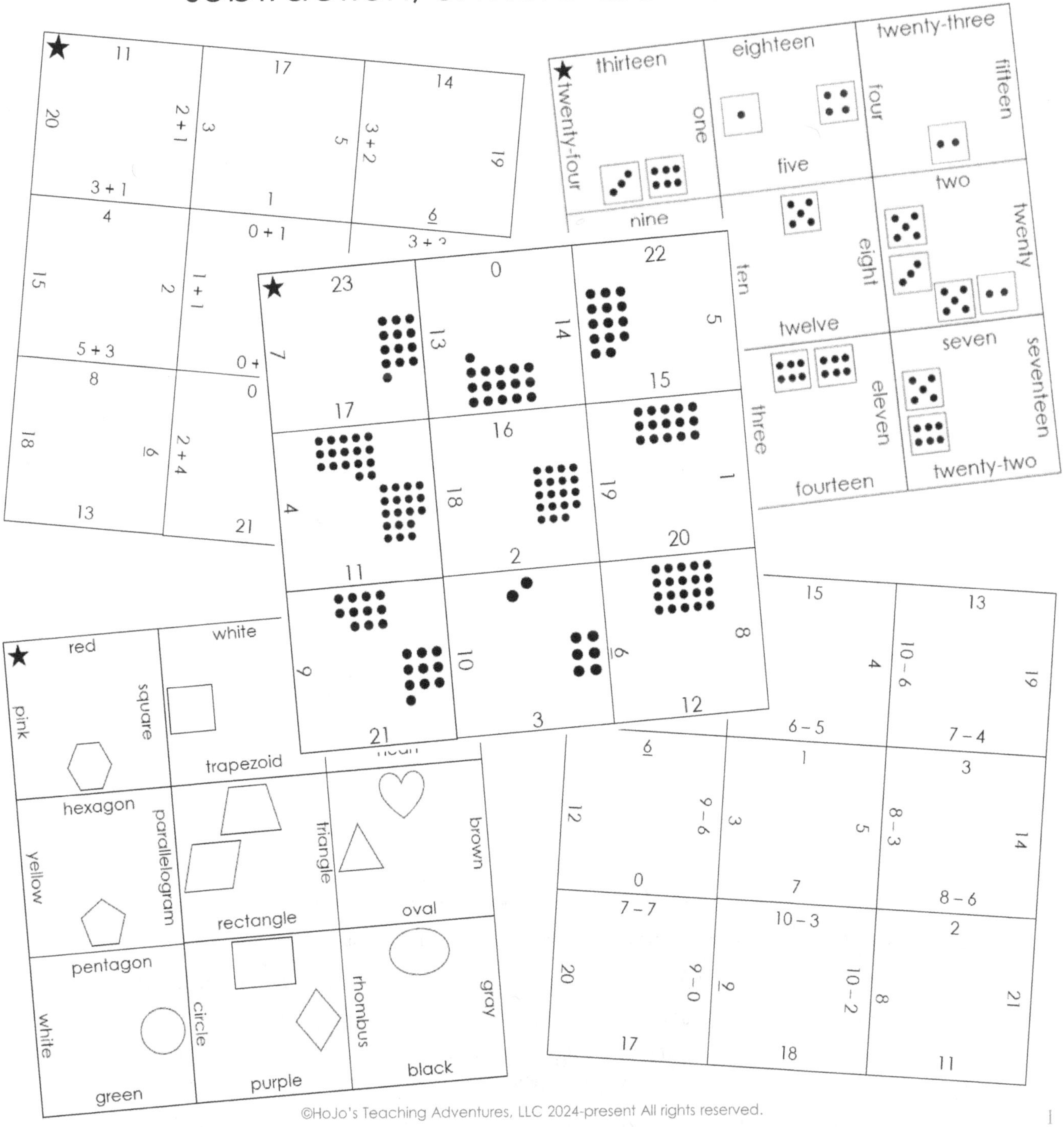

Table of Contents	Page
Matching Number Words 1-12 to Dice	7
Matching Numbers 1-12 to Dice	9
Matching Numbers 1-12 to Words	11
Matching Numbers 12-24 to Words	13
Matching Numbers 0-20 to Place Value Blocks	15, 17, 19
Matching Numbers 0-20 to Ordered Dots	21, 23, 25, 27
Matching Numbers 0-20 to Random Dots	29, 31, 33, 35
Matching Numbers 1-12 to Tally Marks	37
Matching Number Words 1-12 to Tally Marks	39
Matching 2D Shapes to the Word	41
Matching Geometric Shapes to the Word	43
Matching Tens Words 10-120 to Corresponding Number	45
Adding Within 5	47
Subtracting Within 5	49
Adding Within 10	51, 53, 55
Subtracting within 10	57, 59, 61
3x3 Puzzle Mat	63
Answer Key	65-69

Instructions for Use

These puzzles were created to help Kindergarten students practice various math skills in a **hands-on, engaging,** and **FUN** format!

Choose the puzzle you want, cut on the dotted line, cut out the nine puzzle pieces, and hand it to the child. The child will put the 9-piece puzzle back together.

If you want a bit more support for the child, also give them the puzzle mat so they can see the square shape and where the first puzzle piece with the star goes. The puzzle mat is <u>optional</u>, but it does offer more support for kids who are new to these puzzles or those who need a bit more help as they work through the math skills.

There are **28 different math puzzles** to choose from, so you're sure to have a variety of options to work on Kindergarten math all year long!

<u>Ideas for Extra Support:</u>
- Explain to the child that the puzzle piece with the star goes in the upper left-hand corner.
- Make sure they use the included puzzle mat! (The star piece will be easily identifiable this way.)
- Work through the puzzles <u>with</u> the child the first time.

<u>Extended Learning:</u>
- Save the puzzles and let students complete them more than one time.
- Have students create their own puzzle using a square shape with nine pieces. Cut and you'll have another puzzle to put together.

Answer keys for all 28 puzzles have been included at the end of this book.

I hope you enjoy these puzzles as much as I have enjoyed creating them! This teacher turned mom THANKS YOU for your interest in my work! Happy learning!!

★ twenty-four	thirteen	eighteen		twenty-three	
		one		four	fifteen

thirteen / **eighteen** / **twenty-three**

one / five / fifteen / four

| nine | | | two | twenty |
| sixteen | | ten | twelve | eight |

six / ten / twelve / eight / two / twenty

| twenty-one | six | three | fourteen | eleven | seven / seventeen |
| nineteen | | | | | twenty-two |

nineteen / fourteen / twenty-two / three / eleven / seven / seventeen

★ 17 ⟨20⟩ 5	23 ⟨10⟩ 1	13 ⟨11⟩ 18 <u>6</u>
5 ⟨14⟩ 3	1 ⟨12⟩	<u>6</u> 4 ⟨21⟩
3 ⟨19⟩ 16	<u>9</u> 7 22	8 2 24 15

★ twenty-four / 17 / 1 / nine	15 / one / 10 / 8	eighteen / ten / 20 / 2
9 / twenty-one / eleven / five	eight / 11 / four / 12	two / 4 / sixteen / 7
5 / 14 / three / twenty-three	twelve / 3 / six / 19	seven / 6 / 22 / thirteen

Cut here ✂

★ four 7 thirteen 21	12 13 23 seventeen	one twenty-three 9 15
twenty-one two 19 fourteen	17 nineteen twenty-two 24	fifteen 22 six 20
14 11 eighteen five	twenty-four 18 sixteen 8	twenty 16 10 three

★ 13   22   17

24   21

1   4   2

5   6   3

16   7   14

10   8

19   9   12

11   15   18

23   20

★ 1

8

15

7

11

12

3

9

13

5

20

14

16

19

2

17

18

23

4

0

22

10

21

6

★

0

13

14

16

3

7

5

2

9

4

8

10

17

18

12

6

15

21

19

1

20

11

23

22

Cut here

★ 0 1 21 (●●● / ●●)	19 4 (● ... ●●)	15 2 18 (●●●)
5 14 10 (●●●●●●)	6 (●● / ●●) (●●● / ●●● / ●●●)	3 7 13 8 (●●/●●/●●/●)
22 17 (●●●/●●●/●●●)	9 11 23	20 12 16 (●●●● / ●●●●)

★ 23

0

22

7

13

14

5

17

15

4

16

18

1

11

2

19

20

9

10

8

6

21

3

12

★ 15

6

19

22

4

8

1

(●● / ●●)		

14

11

3

17

12

23

0

9

18

5

16

13

20

21

7

2

10

9

Cut here ✂

★ 14 20 7 5	21 9	0 12 23 15
 10 13 2	 1 4	 16 6
 17 8 11	 19 3	 18 22

★ 16 20 1	23 4	13 2 19
5 14 10	6 7	3 17 8
22 11 18	9 0	12 21 15

★ 22

7

0

3

13

14

21

17

15

8

16

19

18

6

1

5

20

23

11

12

9

2

10

4

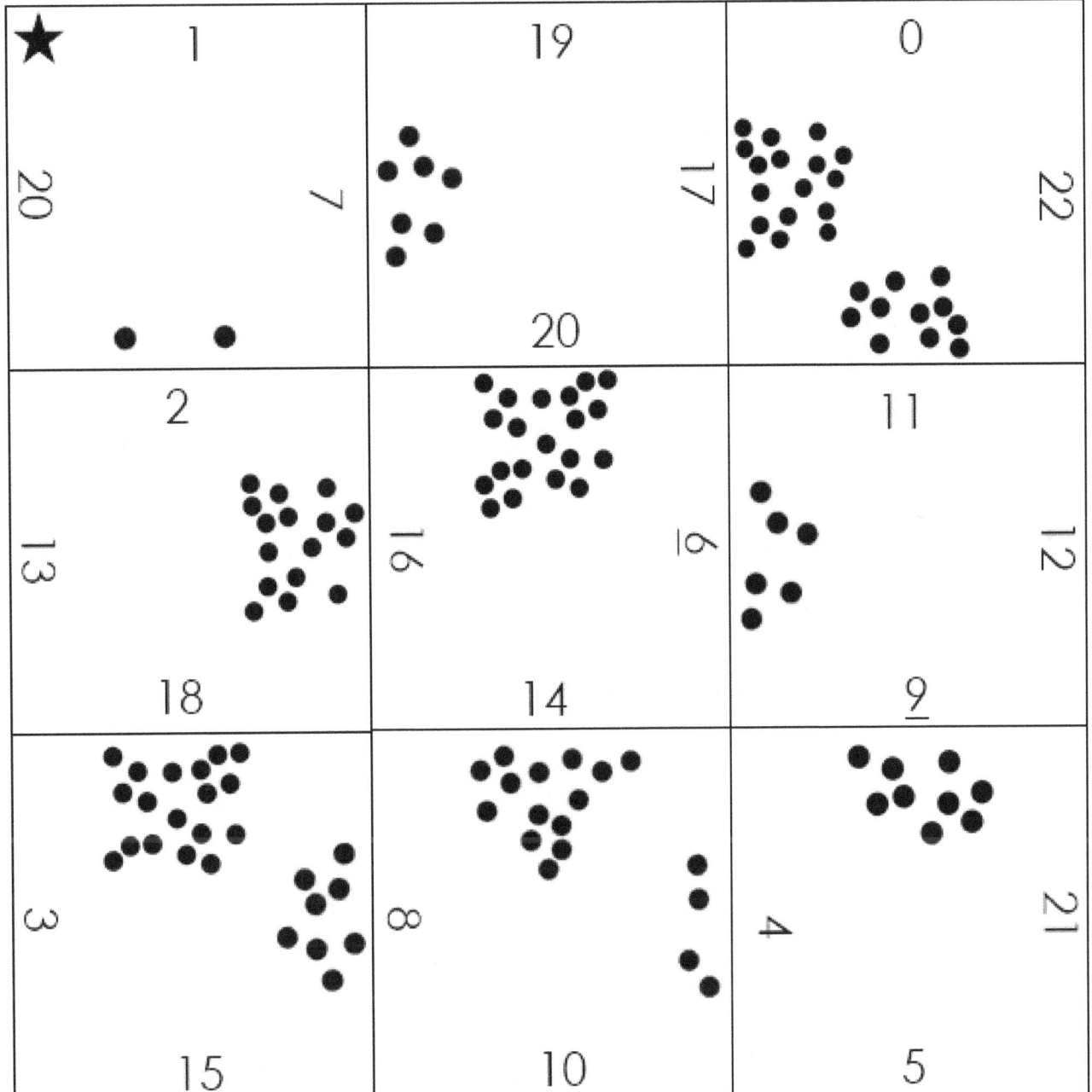

Cut here ✂

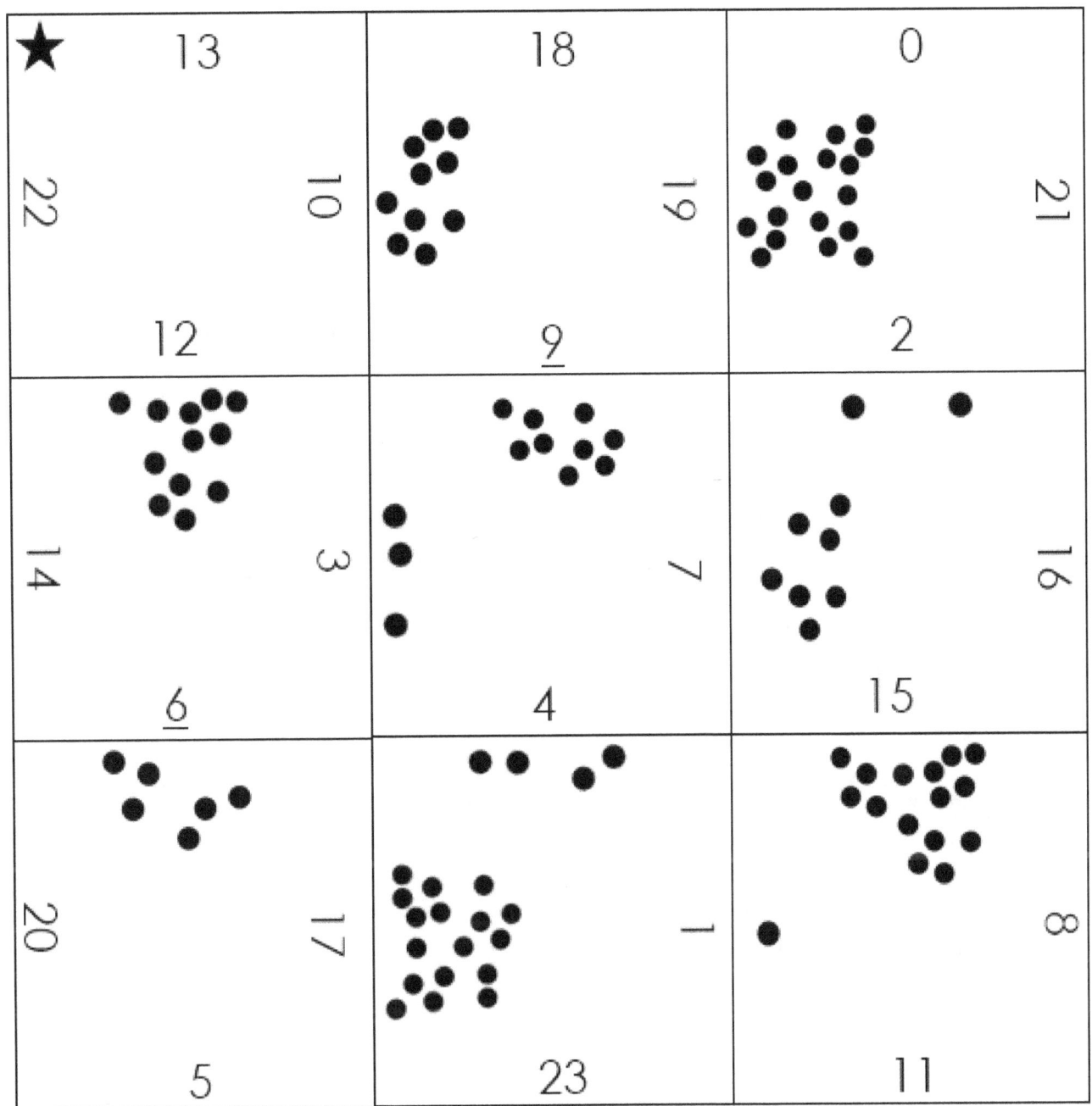

Cut here

★ fifteen	twenty-four	eighteen
twenty · ‖‖‖ · four	eleven · ‖‖‖ ‖‖‖‖ ‖ · one	ten · ‖‖‖ ‖‖‖ ‖ · twenty-one
		seven
seventeen · ‖‖‖ ‖ · six	eight · ‖‖‖ ‖‖‖ ‖ = · twelve	‖‖‖ ‖ ‖ · two · thirteen
		nine
twenty-three · ‖‖‖ ‖ · fourteen	five · ‖‖‖ ‖ · nineteen	‖‖‖ ‖ ‖‖‖‖ · three = · sixteen · twenty-two

red ★	white	blue
pink / square	octagon / trapezoid	orange / heart
hexagon / yellow	parallelogram / rectangle	triangle / brown / oval
pentagon / white	circle / green / purple	rhombus / gray / black

★ pentagon

parallelogram

sphere

square

triangular pyramid

triangle

pentagonal prism

oval

hemisphere

trapezoid

triangular prism

rectangular prism

cone

right pyramid

rectangle

rectangular prism

circle

octagon

cube

hexagon

cylinder

right triangular prism

diamond

rhombus

★ six / thirteen · 30 / sixty	one / thirty · seventy / 40	eleven / 70 · eight / ten
60 / two · one hundred / 90	forty / 100 · one hundred twenty / 50	10 / 120 · five / one hundred ten
ninety / seven · 20 / twelve	fifty / twenty · 80 / four	110 / eighty · three / nine

Cut here ✂

★ 6 12 0 + 1 4 + 1	16 1 0 + 3 0 + 5	9 3 14 1 + 2
5 8 0 + 4 5	5 4 1 + 1 2	3 2 7 3 + 1
3 + 2 15 2 + 2 13	0 + 2 4 5 10	4 5 2 + 3 11 17

★ 6 12 3	16 2 − 1 1 5	9 5 − 5 0 2
3 − 0 8 5 − 2 3 − 3	5 − 0 3 4 4 − 0	4 − 2 5 − 1 7 1
0 15 0 − 0 13	4 0 2 − 0 10	4 − 3 2 11 17

★ 19 / 15 / 1 / 0 + 3	13 / 0 + 1 / 7 / 0 + 10	17 / 0 + 7 / 0 / 0 + 2
3 / 12 / 0 + 6 / 9	10 / 6 / 4 / 0 + 8	2 / 0 + 4 / 16 / 5
0 + 9 / 20 / 5 / 14	8 / 0 + 5 / 2 / 18	0 + 5 / 0 + 2 / 11 / 21

★ 11	17	14
20 ... 2 + 1 ... 3	3 ... 5	3 + 2 ... 19
3 + 1	1	<u>6</u>
4	0 + 1	3 + 3
15 ... 2	1 + 1 ... <u>9</u>	5 + 4 ... 12
5 + 3	0 + 0	2 + 5
8	0	7
18 ... <u>6</u>	2 + 4 ... 10	5 + 5 ... 22
13	21	16

Cut here ✂

★ 20 11 2+8 2 + 2	14 10 2 3 + 4	18 0 + 2 12 7 + 2
4 16 6 + 3 1	7 9 8 6	9 4 + 4 15 3 + 7
1 + 0 21 1 + 2 13	4 + 2 3 4 + 1 17	10 5 19 0

★ 13	20	14
17 1	1 – 0 3	3 – 0 18
6 – 0	8 – 0	5 – 0
6	8	5
22 9 – 0	9 4	4 – 0 12
2	10 – 0	0
2 – 0	10	0 – 0
19 7	7 – 0 2	2 – 0 16
11	15	21

Cut here

★ 17 12 4 – 2 5 – 2	22 2 0 9	15 1 – 1 19 1
3 16 0 9 – 2	10 – 1 10 – 0 4 7 – 1	3 – 2 5 – 1 11 5 – 0
7 21 1 18	6 10 – 9 10 14	5 10 – 0 13 20

★ 22 / 16 / 6 – 4 / 8 – 2	15 / 2 / 4 / 6 – 5	13 / 10 – 6 / 19 / 7 – 4
6 / 12 / 9 – 6 / 0	1 / 3 / 5 / 7	3 / 8 – 3 / 14 / 8 – 6
7 – 7 / 20 / 9 – 0 / 17	10 – 3 / 9 / 10 – 2 / 18	2 / 8 / 21 / 11

Cut here ✂

Put the puzzle together in the correct order.

ANSWER KEY – Page 1

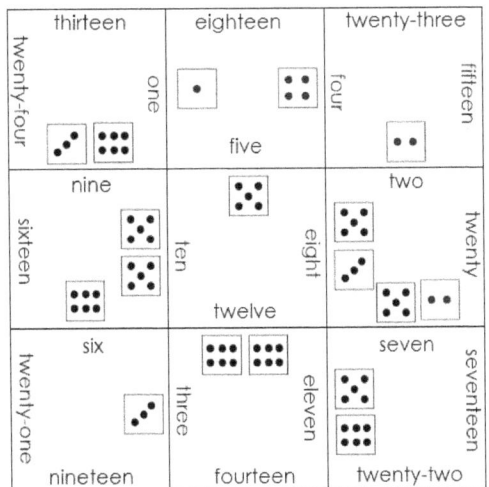

Page 7

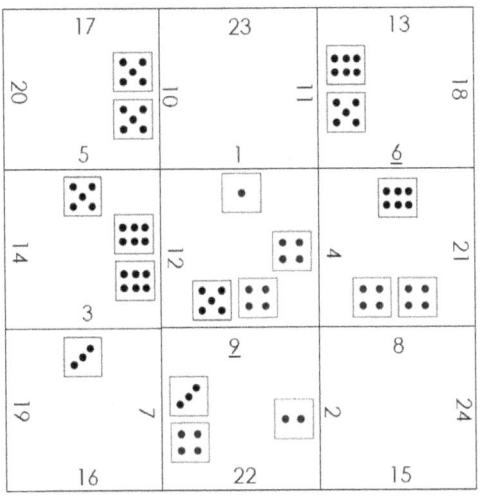

Page 9

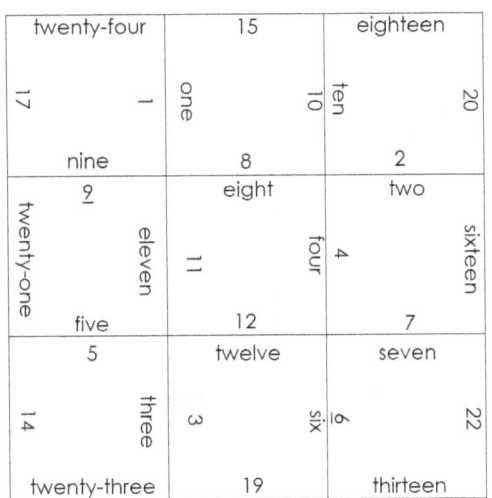

Page 11

Page 13

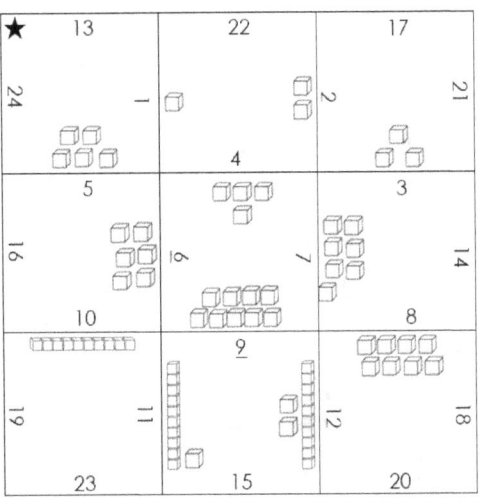

Page 15

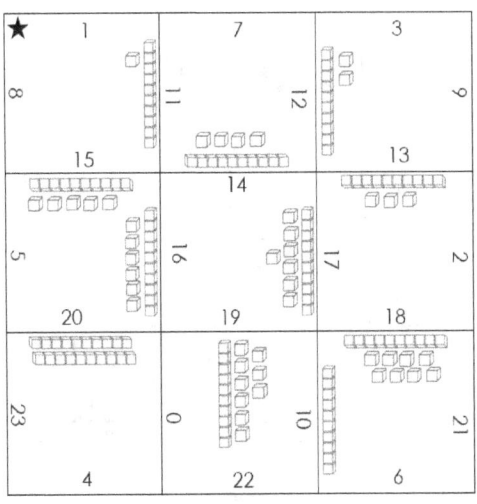

Page 17

ANSWER KEY – Page 2

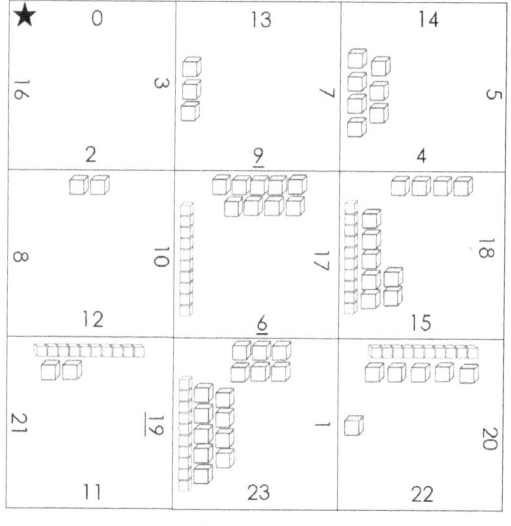

Page 19

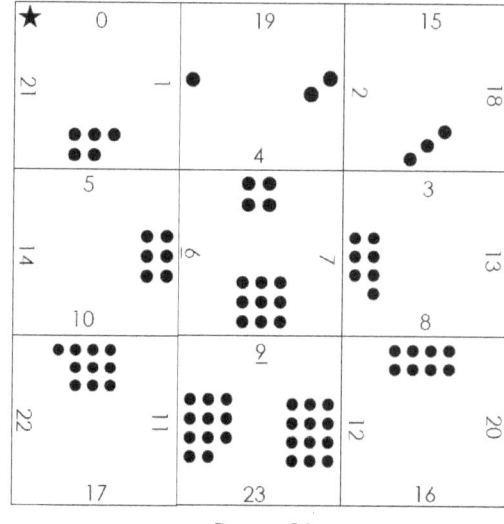

Page 21

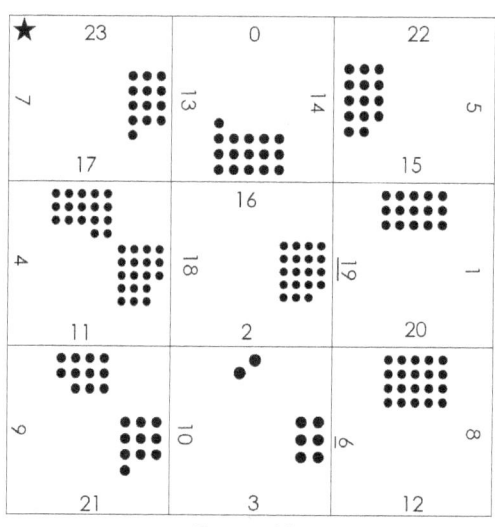

Page 23

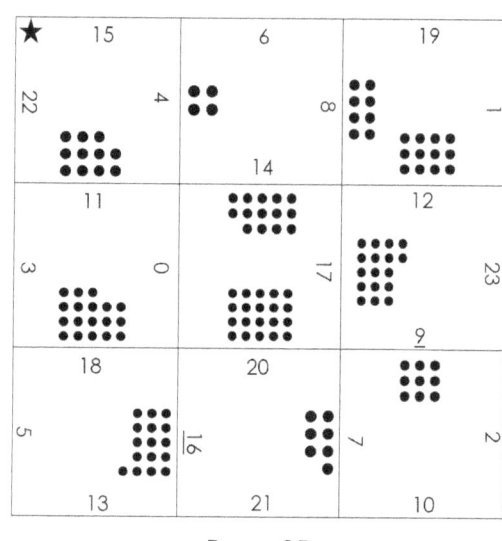

Page 25

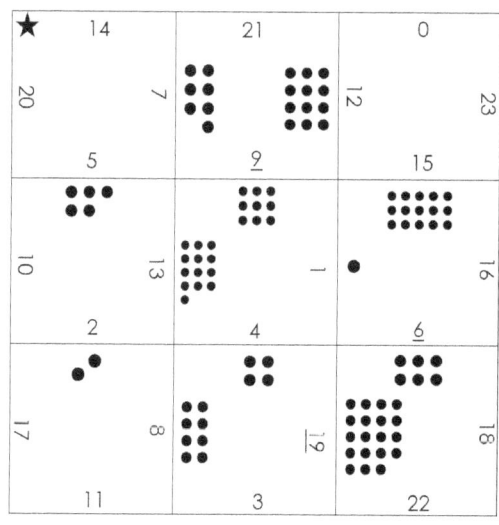

Page 27

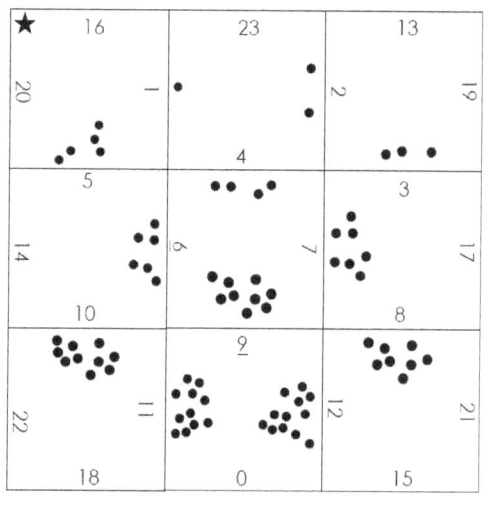

Page 29

ANSWER KEY – Page 3

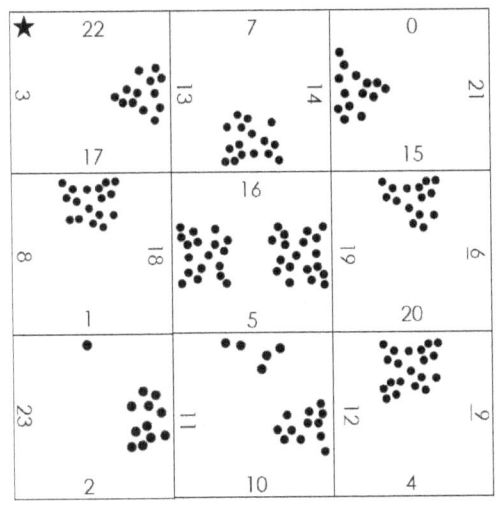

Page 31

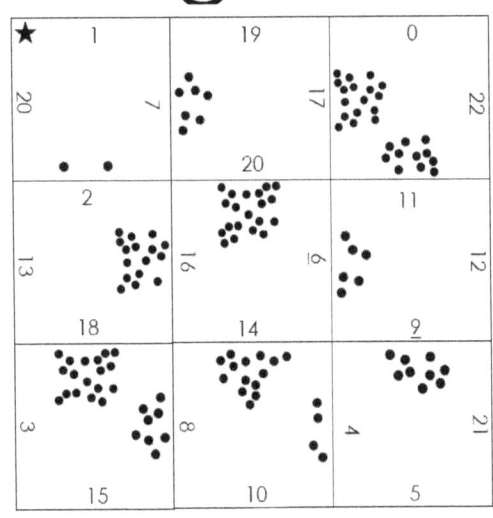

Page 33

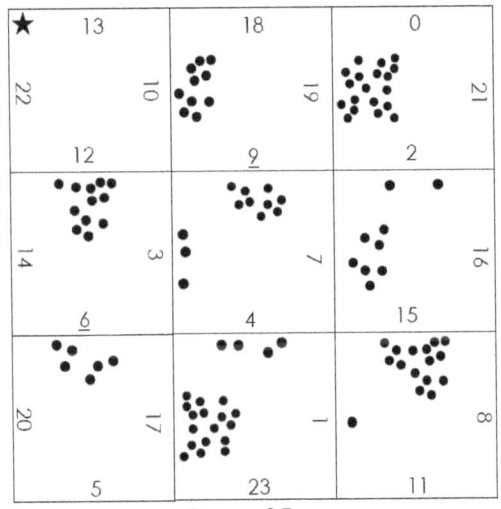

Page 35

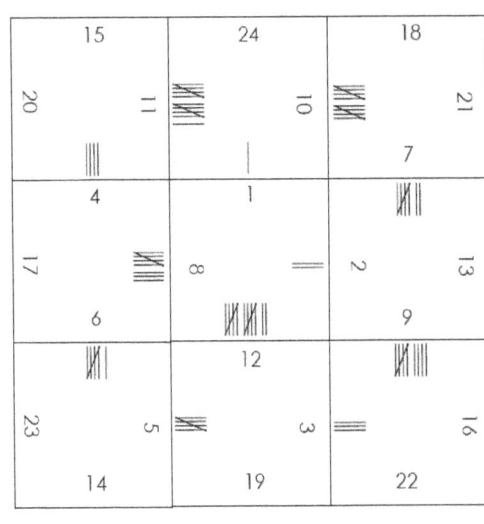

Page 37

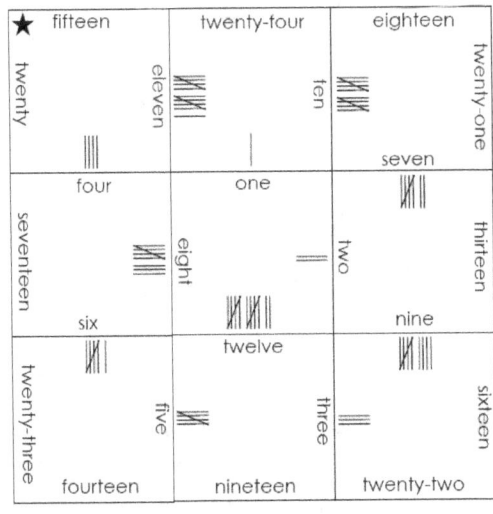

Page 39

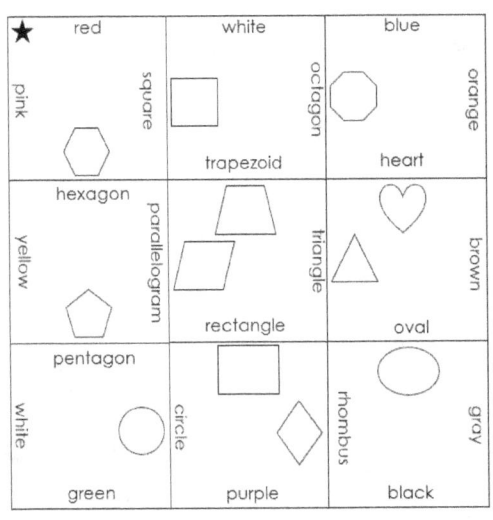

Page 41

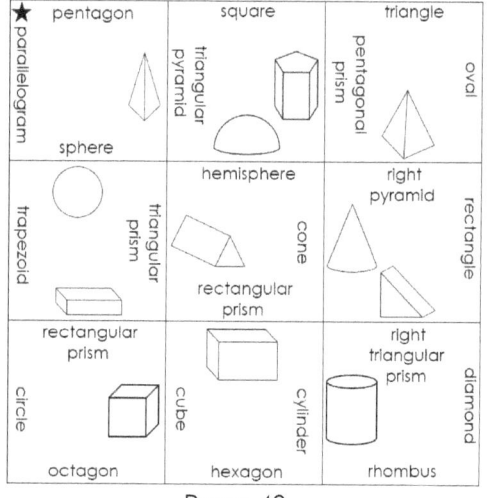

Page 43

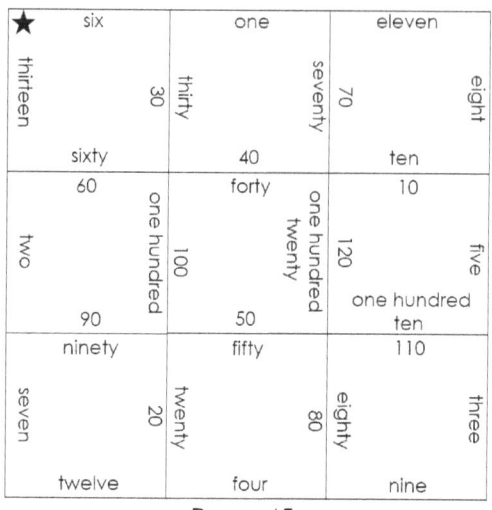

Page 45

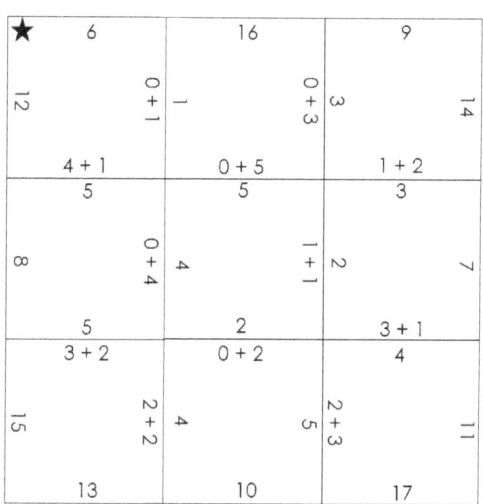

Page 47

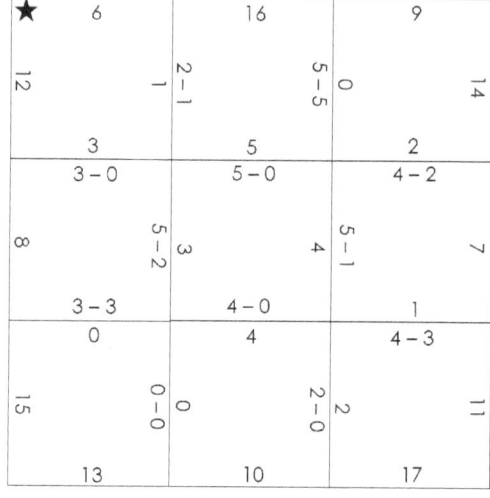

Page 49

Page 51

Page 53

ANSWER KEY – Page 5

Page 55

★ 20 | 14 | 18
2 + 8 (10) · 2 + 2 (11) — 2 (10) · 3 + 4 (10) — 0 + 2 (0+2) · 12 · 7 + 2

4 | 7 | 9
6 + 3 · 2 (2) · 16 — 8 · 4 + 4 · 15 — 1 · 6 · 3 + 7

1 + 0 | 4 + 2 | 10
21 — 1 + 2 · 3 — 4 + 1 · 5 · 19 — 13 · 17 · 0

Page 57

★ 13 | 20 | 14
17 — 1 · 1 – 0 · 3 · 3 – 0 · 18 — 6 – 0 · 8 – 0 · 5 – 0

6 | 8 | 5
22 — 9 – 0 · 2 · 4 – 0 · 12 — 2 · 10 – 0 · 0

2 – 0 | 10 | 0 – 0
19 — 7 – 0 · 7 · 2 – 0 · 2 · 16 — 11 · 15 · 21

Page 59

★ 17 | 22 | 15
4 – 2 · 12 — 2 · 2 · 1 – 1 · 0 · 19 — 5 – 2 · 2 · 1

3 | 10 – 1 | 3 – 2
10 – 0 · 16 — 0 · 4 · 5 – 1 · 11 — 9 – 2 · 7 – 1 · 5 – 0

7 | 6 | 5
10 – 9 · 21 — 1 · 10 · 10 – 0 · 13 — 18 · 14 · 20

Page 61

★ 22 | 15 | 13
6 – 4 · 16 — 2 · 4 · 10 – 6 · 19 — 8 – 2 · 6 – 5 · 7 – 4

6 | 1 | 3
9 – 6 · 12 — 3 · 5 · 8 – 3 · 14 — 0 · 7 · 8 – 6

7 – 7 | 10 – 3 | 2
9 – 0 · 20 — 9 · 10 – 2 · 8 · 21 — 17 · 18 · 11

www.HoJosTeachingAdeventures.com

Want FREE math puzzles you can use today?